The Proteaceae

The PROTEACEAE of SOUTH AFRICA

FRANK ROUSSEAU

PURNELL
CAPE TOWN · JOHANNESBURG · LONDON

PUBLISHED IN SOUTH AFRICA BY
PURNELL AND SONS (S.A.) PTY., LTD., CAPE TOWN

© Frank Rousseau 1970

SBN 360 00106 8
SET IN 11 PT. ON 12 PT. AND 12 PT. ON 13 PT. MONOTYPE BEMBO
PRINTED AND BOUND IN THE REPUBLIC OF SOUTH AFRICA BY
THE RUSTICA PRESS, PTY., LTD., WYNBERG, CAPE
LITHOGRAPHIC POSITIVES BY MESSRS. HIRT & CARTER (PTY.) LTD., CAPE TOWN

Foreword

Only a small number of the species described in this book are to be seen in gardens. If all could be found in gardens my task would have been relatively easy and, no doubt, the photographs would have been more impressive as I could have taken my pictures when conditions were most favourable. As it is, the vast majority of species can only be seen growing wild in their natural habitats, which may be anywhere in the rugged mountains from Grahamstown across the width of the Republic to Cape Town and, from there, northwards as far as Clanwilliam.

It would have been so much easier if they all could be seen flowering on the same mountain at the same time. Although certain species do have a fairly wide distribution in the south-western Cape the majority are to be found only in different, limited localities and some of the rarest species can only be found in almost inaccessible spots high in the mountains. Neither do they all flower at the same time of the year. Although most proteas flower during the colder months, for any month of the year I could name species which are at their best at that particular time.

Proteas like wind and the majority grow on the south-facing slopes exposed to the cool south-east winds from the sea. This does not make photography easy. It takes several hours to climb a mountain and to find the species you are seeking. The light is hard when the South African sun is high and, because of the hard shadows and the wind, I was sometimes obliged to spend the night in the mountains in the hope that early morning would bring more favourable conditions. On many occasions I had no choice, and some of my photographs were taken in rain and some even in mist. Sometimes I failed to find the species I was seeking, and sometimes I found it only to discover that I was either too early or too late to see it in flower. My disappointment was acute when I found that I was too late as it meant that I would have to wait a whole year before the opportunity would again present itself.

It took five years before I was reasonably satisfied with the completeness of my collection and the standard of my photographs, and now I am glad that I could not take most of my photographs in gardens and was obliged to study the species growing wild in their natural environments. Proteas hybridize easily in gardens and you can never be sure that any specimen is pure unless you have seen it in its wild state.

Why go to all this trouble? In reply I can do no better than quote Edmund Hillary when he was asked why he wanted to climb Mount Everest—'Because it is there'.

It was a challenge.

The proteas are there, but some species have already become extinct and others are doomed to extinction. All are fated to become more rare. On the plains they have had to make way for crops and pastures, and in the mountains they are threatened by fires. Our country is short of timber and, wherever the mountain soil is deep enough, pine trees are being cultivated—and nothing will grow under pines. Moreover they create a fire hazard and each summer serious fires break out in the mountains and thousands of acres of our wonderful flora are destroyed. Hardly any part of these mountains near the Cape has remained unscathed.

An occasional light and restricted fire is actually beneficial in that it clears away tall, old, and woody shrub and dense undergrowth, so giving seedlings scope to establish themselves. Fires have always occurred occasionally and sporadically in Nature—usually lightning sets fire to the most dry vegetation and the downpour which usually follows limits its spread. The coming of man has altered the pattern and species are exterminated only when fire is repeated too often and too soon in the same area.

Fire is also an ally of Hakea. This is a distant relative of the proteas which was introduced from Australia as a suitable hedge-plant. Now our mountains are infested with it and the prickly shrubs grow so densely that they choke all other vegetation. It is not destroyed by fire. On the contrary, fire only causes the hard 'cones' to split open and release small, winged seeds which are scattered far and wide by the wind. Where the natural vegetation has been razed by fire these seeds come up as thickly as the proverbial hairs on a hog's back. No effective means has yet been found of destroying Hakea, but the introduction of their natural enemies—certain insects—is being investigated (*Hakea sericea*—see page 102).

Invaluable work is done at the National Botanic Gardens at Kirstenbosch by cultivating and preserving our indigenous flora, and several species of Proteaceae which have become almost extinct in Nature can be seen here, but many of the rarest species have not responded to cultivation and evidently need the crisp air and the cool mists of the high mountains.

I wanted a complete photographic record of all the South African Proteaceae. Since there are close to 400 distinct species I soon realized that I might never achieve this aim, but at least I could try. I thought of the slogan, 'Go as far as you can, then see how far you can go'.

What I had in mind was a book intended for the general public in which I could show the wonder of our Proteaceae in all their amazing variety and, naturally, I wanted this book to be a complete record. That such a book would ever appear seemed highly unlikely to me. In view of the high cost of colour printing it would be impractical.

My search for the proteas also incurred expense, and this I recouped by writing articles, illustrated with photographs, for our leading magazines. My articles on proteas appeared in *Personality* and *South African Panorama* but first, and most important, was a series on the Genus Protea which appeared weekly for a period of six months in *Die Huisgenoot*. The response was most encouraging and there were many letters of appreciation from the readers. It became apparent that there was keen interest in our Proteaceae and that there was a need for a book such as this.

The publishers were even more hopeful than I was and, after calculating costs, I was asked whether I could cover the subject in 100 colour pages. I felt that I could. Although the book would not be complete, and many species would have to be omitted, all were not sufficiently interesting or attractive to warrant the expense of inclusion. Others were sufficiently similar for a brief comparison to suffice.

ACKNOWLEDGEMENTS

I am deeply grateful to the following authorities, not only for the information which they so generously and willingly supplied, but also for their encouragement:

MRS. MARIE MURRAY VOGTS
MR. ION WILLIAMS
PROFESSOR BRIAN RYCROFT
MR. JOHN ROURKE

Leisure Island,
KNYSNA.
1970

Contents

	Page
FOREWORD	v
Acknowledgements	vii
INTRODUCTION	xi
Terminology	xi
The Hypothesis of Wegener	xii
Proteas and evolution	xiii
Nomenclature	xvi
THE FAMILY, PROTEACEAE	1
THE PROTEACEAE OF SOUTH AFRICA	1
THE GENUS PROTEA	2
Sections with main stem overground	3
Section Cynaroideae	3
Section Melliferae	6
Section Ligulatae	8
Section Speciosae	12
Section Exsertae	26
Section Pinifoliae	30
Proteas of the summer-rainfall area	39
Proteas with main stem underground	40
THE GENUS OROTHAMNUS	47
THE GENUS LEUCOSPERMUM	48
Leucospermums with large flowerheads	48
Leucospermums with small flowerheads	56
THE GENUS MIMETES	60
THE GENUS SERRURIA	63
THE GENUS PARANOMUS	66
THE GENUS AULAX	68

	Page
THE GENUS BRABEIUM	70
THE GENUS LEUCADENDRON	71
THE GENUS SPATALLA (SPATALLOPSIS)	96
THE GENUS SOROCEPHALUS	96
THE GENUS DIASTELLA	97
THE GENUS FAUREA	97
HYBRIDS	98
IN THE MOUNTAINS	100
EXAMPLES OF AUSTRALIAN PROTEACEAE	102
HOW TO GROW PROTEAS	103
PROTEAS LOST, AND FOUND AGAIN	105
COMMON NAMES	107
INDEX	109

Introduction

All plants of which the flower has a perianth of 4 segments, and 4 stamens opposite the periant-lobes, situated in spoon-like depressions of the lobes, with free 2-thecous anthers opening lengthwise, and a simple style from a 1-chambered superior ovary, belong to the family, Proteaceae.

There is no need to feel discouraged if you do not understand the above definition. Although I hope that the botanist will find much of interest in these pages, this book was written primarily for the general public and is not intended as a scientific treatise on the subject.

The parts of the individual flower of a protea are shown in the following sketch.

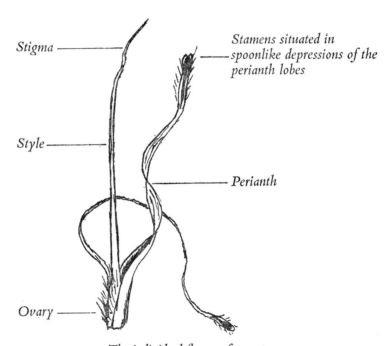

The individual flower of a protea.
(In the Genus Protea one segment of the tubular perianth becomes separated from the other three fused lobes when the flower opens.)

In my brief descriptions of the various species I have avoided complicated terminology and have used only a few elementary botanical terms. These can easily be explained.

A protea has a *composite flower*, which means that it is composed of many small, individual flowers. In a flower the *bracts* are those green, leaf-like structures which enclose the young buds. In a protea these bracts are large and become brightly coloured so as to simulate petals both in appearance and function. The composite flower, together with these surrounding coloured bracts, is spoken of as the *flowerhead* or the *inflorescence*. In the genera Leucadendron and Mimetes, the leaves surrounding the flowers at the ends of the branches may also become coloured and in this book, for the sake of clarity and simplicity, I have confined my use of the term *inflorescence* to this whole coloured structure, but *flowerhead* could also have been applied here with equal correctness.

In the closed flower the smooth style is ensheathed by the perianth, which usually has a coating of silky hairs. The stiff style flexes in order to pull itself out of this sheath, the stigma being the last to be released. The perianth of the open flower is now limp and flaccid and falls to the bottom of the 'cup'. In a 'Pincushion' (Leucospermum) the styles are the 'pins' and you will notice that the perianths are rolled back towards the 'cushion' or central 'hub'.

The south-western Cape has so many wonderful species of wild flowers that it has been described as the richest botanical treasure-store in the world—there are more than 2 500 distinct species in the Cape Peninsula alone—and the one that attracts most attention is the protea.

Yet many are disappointed when they first see a protea and find it too large and stiff. Proteas do not wilt and can be kept in the vase for weeks. They merely dry out gradually and lose their colour. This may be a disadvantage, since the flowerhead you first see might not be fresh and it might not have been picked at its most attractive stage. Most proteas do not open wider after being picked.

Proteas may not be pretty in the sense that roses and carnations are pretty, but they are always striking. They are proud, and dignified, and magnificent, and when you go to a wild-flower show in the Boland and see one of those massive protea arrangements at which the ladies there are so expert, then you will be won over and the protea will begin to cast its spell over you.

But proteas are not always large, and stiff, and proud. In this book you will see many species which have small and dainty flowerheads. It is this amazing variety of forms which fascinates me most.

THE HYPOTHESIS OF WEGENER

The largest number of species of the Family Proteaceae is found in South Africa and in Australia. This is surprising since Australia is situated on the other side of the globe and there are thousands of miles of ocean between the two continents. The only explanation which naturalists will accept for this relationship is the Hypothesis of Wegener of the Drifting Continents. This has since been amended by Barnett. He cut out the continents on a plastic globe and then, by fitting them around a globe half as big, demonstrated that the continents interlock as neatly as the pieces of a jigsaw puzzle. His theory is that, 3 000 million years ago, the earth was half its present size and then expanded at the rate of 2 inches per 100 years until, 500 million years ago, large cracks appeared in the earth's crust. As the earth continued to expand these cracks widened and so the continents were formed. Much geological evidence has accumulated which lends support to this hypothesis. For example; if the continents could be brought into proximity certain geological strata can be traced from one to the other; and the magnetic orientation of certain rock formations with the poles indicates that such a shift did indeed take place.

PROTEAS AND EVOLUTION

In most of the Australian Proteaceae the flowers are arranged in groups on spikes so that the flowerhead is of the type popularly known as a 'bottle-brush', and the only one which is easily recognized by the layman as being a relative of our protea is the 'Waratah' (*Telopia speciosissima*), which also has coloured bracts surrounding the composite flower such as we find in a protea (page 102).

I find this very interesting considering that their common ancestor must have existed so very long ago in the distant past. You have only to look at the Proteaceae illustrated in this book to realize what great changes have been brought about over the centuries by evolution.

A plant family includes all those members which have certain structural characteristics in common, indicating that they are related, and it is evident that there has been some link in their evolution.

The family is divided into genera and each genus includes those species which are so closely related that they are capable of crossbreeding with each other. The result is known as a hybrid—but you cannot cross a species of one genus with the species of another. The gap is too large.

Exactly how this gap was bridged and how a new genus evolved we do not know. It is thought that a new genus is the result of a remarkable mutation. A mutation is the sudden and unexpected appearance of a freak whose deformity was not inherited from its parents but which may be passed on to its offspring. Mutations are not uncommon in Nature and occur more frequently at times when the earth is bombarded by cosmic rays. They are almost always a handicap and the progeny will eventually die out. Should, however, the rare exception occur and the mutation is advantageous then, so it is thought, this might be the beginning of a new genus.

It is evident that hybridization plays a negligible role in evolution. A hybrid is seldom an improvement and the weaklings have a tendency to die out in Nature. Although it does occasionally happen that a hybrid is an improvement and succeeds in establishing itself, such short-cuts in evolution are exceptional and, if left alone, the descendants of the hybrid will eventually revert back to type.

No two individuals in the Plant or Animal Kingdoms are exactly identical. Each has its own minor characteristics which are hereditary and which may be passed on to the offspring. Those who have inherited characteristics which are advantageous, or favour their adaptation to a slowly but constantly changing environment, have the best chance of survival and are the ones who will pass these characteristics on to future generations. Those who have inherited unfavourable characteristics will eventually die out, even though it may take many successive generations. Evolution is a slow, continuous process of either adaptation or elimination. It is a process of continual experimentation in which new species are formed and others are discarded.

This might take thousands of years. Nature is in no hurry and has all the time there is. Evolution is too slow a process for man to observe in his ordinary life-span, and proteas fascinate me because here I find sufficient evidence of the process of evolution, on which I may speculate. Evolution does not follow a straight line and its ramifications are so intricate that we shall never be able to unravel the pattern.

Only very occasionally is the process of evolution direct and rapid enough for us to perceive exactly what has taken place, and the best example I can think of is that of the peppered moth. This moth is found near Manchester and during the day it settles on the bark of trees where it is so well camouflaged that it escapes the notice of its natural enemies—birds. Early specimens and descriptions show this moth to be grey in colour with brown-speckled wings, exactly matching the bark of the trees. But then came the industrial revolution and with the smoke and smog of Manchester the bark of the trees began to blacken. What happened to the moth? Since no two moths are exactly identical the darker ones survived but the majority, which now appeared lighter than the bark of the trees, became conspicuous and were gobbled up by the birds. As the bark of the trees became darker so only the darkest moths escaped and lived to carry on the propagation of their kind. Today the grey peppered moth is black.

In the case of a plant, a change of environment might be a matter of only a few yards, and in the proteas we find not only different species which are very similar in appearance, but the same species may show minor variations from one locality to another. These local variations we call 'forms'. I have included a few of the forms in this book because I find them so interesting. I know that these are the experiments of evolution, but I can only speculate on how or why they evolved.

For instance, at Knysna on the lush and more shaded south-facing mountain slopes, *Protea cynaroides* has the usual, large, oval-shaped leaves, but only a few miles lower on the plain near the coastline the same species has small, narrow leaves. It is hotter and drier here and a plant would lose more moisture through transpiration, and it seems feasible to me that the individuals who succeeded in establishing themselves here were the ones who, by chance, happened to have smaller leaves. This might be why and how it happened, and eventually this might lead to a new species, but I am afraid I shall never know the answer. It seems as likely that the experiment will abort, especially now that more and more plantations are being established here, and these same small-leafed proteas can now be seen stretching towards the light. Man has changed their environment too rapidly and they won't have sufficient time to readapt.

Why do several species of protea have long hairs at the tips of their bracts? There must surely be some reason and these hairs must in some way serve a purpose.

The sugar-bird uses the dry flowers of the protea when he builds his nest and inevitably a few seeds are incorporated. Later the old nest will be blown about by the wind and these seeds will be scattered. It seems reasonable to me that the sugar-bird would prefer a dry flowerhead which looked more woolly, and it seems logical that, when first occasional proteas appeared which accidentally happened to have long hairs at the tips of their bracts, these were the ones which were preferred, and had their seeds scattered by the birds, giving them a greater chance to germinate. Through subsequent interbreeding this characteristic would have been retained and developed. This is how it might have happened.

Most of our popular garden flowers are the result of man's interference with Nature and are hybrids, but the horticulturists have not yet begun to experiment with our Proteaceae and, except for the occasional accidental hybrid, the species you see growing wild are the creations and experiments of natural evolution.

In order that they might easily be compared I have arranged my photographs so that species which look fairly similar appear in sequence. I arranged the species of the Genus Serruria in a sequence according to the size of the bracts, and then I noticed that those species which had the largest bracts were the ones which were most rare and those whose bracts were insignificant were the common ones. Does this mean that the experiment with large bracts was not a success and that these species are being discarded? I doubt it, but I wish I knew!

Also, so as to have some semblance of order in the section on the Genus Protea, I have adopted the classification of Phillips as modified by Marie Murray Vogts. Mrs. Vogts is the first to admit that this classification is not altogether satisfactory, and that the delineation between the sections is not always clear-cut, but I have followed this classification because I think it is the best we have.

I have always regretted that, in scientific literature, there is a tendency to carry down mistakes or misrepresentations contained in the original publications. For this reason I have not referred to Phillips's *Flora Capensis,* and my descriptions of the various species are based on my own observations.

NOMENCLATURE

The only valid name is the one given in the first published description. The early explorers sent back specimens to scientists in their home countries in England and Europe. Seeing that the first descriptions were consequently made from dried specimens and the records preserved in the various archives overseas, it is not surprising that botanists later named species which had already been described, either not recognizing the plant from the earlier description, or through being quite ignorant of the fact that an earlier description existed.

The various genera are at present being revised by different authorities and, as the archives bring to light these mistakes, so the names have been changed back to the valid ones.

Although I have used the valid names in this book, and also give the old ones where changes have been made, the revision will still take several years to complete and there are likely to be further changes in the future.

The Family Proteaceae

This is a large family with 61 genera and about 1 500 species of which the majority are found in Australia and South Africa, but it is fairly widely distributed throughout the southern hemisphere and the remaining species are found in South America, tropical Africa, Malaya, New Zealand and the Pacific islands. All are evergreen shrubs or trees.

THE PROTEACEAE OF SOUTH AFRICA

The South African family consists of 13 genera and close to 400 species.
None of these is found in any other continent and, except for certain species in the Genus Protea and in the Genus Faurea, which are found as far north as Central Africa, the vast majority are found growing wild only in the south-western Cape.

The South African Proteaceae are divided into the following genera:

PROTEA	BRABEIUM
OROTHAMNUS	LEUCADENDRON
LEUCOSPERMUM	DIASTELLA
MIMETES	SPATALLA
SERRURIA	SOROCEPHALUS
PARANOMUS	FAUREA
AULAX	

Up to this year a fourteenth genus, SPATALLOPSIS, was recognized. In a recently published monograph on SOROCEPHALUS and SPATALLA, Mr. J. P. Rourke of the Compton Herbarium, Kirstenbosch, concluded that SPATALLOPSIS was synonymous with SPATALLA.

The Genus Protea Linnaeus

With more than 100 species this is our largest genus. It has the largest and most colourful flowerheads as the composite flower is surrounded by an attractively shaped whorl of large, brightly coloured bracts.
Although the majority of species is found in the south-western Cape, certain species occur on both sides of the Drakensberg and others are found as far north as Central Africa, but these are in the minority and cannot compare with the beauty and variety of those from the south-western Cape.
They may be divided into two main groups—those with main stem overground, and those with main stem underground.
The name Protea was given by Linnaeus in 1735, and was derived from Proteus, the Greek god who, according to legend, was able to change his shape and appearance at will.
It was from the name *Protea* that the name *Proteaceae* was given to the family by the French botanist, Jussieu.

SECTION CYNAROIDEAE

Protea cynaroides L.

Popularly known as the Giant Protea, but it is sometimes referred to as
the King Protea, this species has the largest flowerhead—up to 30 cm across.
It is found from Cape Town to Grahamstown but is no longer very common in
Nature. The shrub is only about 1 metre in height and the leaves,
which are large, smooth, and leathery, have long leaf-stalks. They are usually oval in
shape but this differs from one locality to another. For example, on
Table Mountain they are round. At the coast near Knysna they are long and
narrow, whereas only a few miles inland towards the mountains they are oval in
shape. The flowering time is either in autumn or spring, but the same plant
does not usually flower in both seasons. On their outsides the bracts have a covering of
very fine silky hairs which gives them an attractive velvet-like sheen.

Protea cynaroides

Although the most common colour is a clear shell-pink, it may be found in any of the shades between white and deep red.
The dark red specimen, shown above, grows near Storms River, where it is rare.

Protea cynaroides (var. glabrata Meisn.)

This is a rare variety from the Tzitzikama. It differs in that the bracts are smooth and not covered with the velvet-like hairs on their outer surfaces. They are longer and more pointed and they do not open as wide. The buds are long and spindle-shaped and unlike other forms of this species which may be globular in shape.

Also of interest is the little dwarf form which is found growing near Port Elizabeth in acid, sandy, wind-swept soil. The plant, the leaves, and the little flowerheads are only half as large as those usually found in *P. cynaroides*.

SECTION MELLIFERAE

Protea repens (L.) L.

Popularly known as the 'sugarbush', this is the Republic's national flower. It is found from Cape Town to Grahamstown, both on the plains and in the mountains, and is still relatively common. The bracts are smooth and shiny, and slightly sticky as the cup contains nectar from which the early colonists brewed a syrup—'suikerbos-stroop'. It flowers in early autumn on an erect shrub about 3 metres high. Thunberg appropriately named it *P. mellifera*, but the name had to be changed when it was discovered that Linnaeus had already named this species.

Protea repens

The white form is more common near the coast and near Cape Town. The deepest colours are found near the Langkloof.

SECTION LIGULATAE

Proteas in this section have spoon-shaped bracts

Protea eximia (Salisb. ex Knight) Fourcade

Well-known under its old name, *P. latifolia*, this protea has a long flowering period from autumn till early summer and, when there is sufficient rain during the dry summer months, it may flower throughout the year. The tree may become as tall as 3 metres. It is found in the mountains on both sides of the Langkloof and Little Karoo. The flowerhead is about 13 cm in length.

Both proteas on this page come from the Bredasdorp area where they grow in slightly alkaline soil, in contrast to the majority of species, which demand acid soil.

P. obtusifolia Buek ex Meisn.

Has smooth and shiny (but not sticky), bright-red bracts, but there is also a less common pure white sort. It flowers in autumn for a long period on a shrub only about 1 metre in height.

Protea susannae Phillips

Can also be seen growing next to the roadside at Albertinia where it flowers in autumn on a tree about 2½ metres in height. The leaves, when crushed, give off a slightly unpleasant odour. The bracts are always reddish-brown in colour.

Protea longifolia Andrews

Grows in the Hermanus area where it flowers in early winter on a shrub about 1 metre in height. The colour of the smooth bracts is always pale apple-green, sometimes with reddish-brown shadings.

Protea rouppelliae Meisn.

Although this protea is found in the mountains of the eastern Cape and further north on both sides of the Drakensberg, it does not have the typical shape of the summer-rainfall proteas and resembles those of the south-western Cape. It is cup-shaped and the ligulate bracts are longer than the flowers. The colours may be either pink or white and the bracts have a fine coating of silky hairs on their outer surfaces. The leaves, of the sturdy 4-metre high tree, are also covered with silky hairs to give them a silvery sheen.

Protea lorifolia (Salis. ex Knight) Fourcade

Is found in the mountains on both sides of the Langkloof where it seems to prefer the inland-facing slopes. The tree is sturdy but only about 1½ metres in height, and the leaves are about 22 cm in length. It flowers in spring and the colour of the bracts may be white, cream, or pink.

Protea compacta R.Br.

Popularly known as the Bot River Protea, it also grows on the other side of Hermanus near Stanford. The slender shrub has tall branches which may reach a height of about 4 metres. The flowerheads are situated at their tips and are beautifully shaped like a wineglass. They are deep pink in colour and slightly translucent so that they become radiant when back-lit by the sun. A white colour is also found but is uncommon. Flowering time is in early winter.

SECTION SPECIOSAE

Proteas in this section have hairs at the tips of their bracts (bearded proteas)

Protea neriifolia R.Br.

This is a fairly common protea and it is widely distributed along the south-western Cape coastal belt. The tree grows to a height of about 3 metres but, as in the case of almost all proteas, the best flowers are found on much younger shrubs. The bracts have a satin-like sheen and, although the most common colour is pink, they may be found in any shade between white and deep red. The purplish-black 'beard' is most attractive. It flowers from autumn to early summer.

Protea laurifolia Thunberg

The flowerhead is almost indistinguishable from that of *P. neriifolia* but the colour is always salmon-pink and the leaves are broader. The tree is much sturdier and may reach a height of 6 metres. It has roughly the same distribution.

Protea lepidocarpodendron (L.) L.

Is found along the coast between Cape Town and Port Elizabeth. It is never very colourful, but attractive nevertheless, with the strong contrast of its black-tipped bracts. It flowers in spring.

Protea speciosa L.

Comes from the high mountains near Hermanus. The brown-bearded bracts are found in colours of white, cream, or pink and have a most attractive satin-like sheen. The upright tree is about 1½ metres in height.

Protea speciosa L. var. *angustata* Meisn.

Also comes from the high mountains near Hermanus and is rare. The bracts open a little wider than shown in the photograph. It is also known as *P. patersonii* L.Bol.

Protea stokoei Phillips

This is a very rare and lovely protea which grows high in the Hottentots-Holland mountains. Attempts to cultivate it lower down have not met with much success as it evidently demands the cool mists which form here when the south-easter blows from the sea in the dry summer months. It has however been successfully cultivated in New Zealand. It is rather like *P. speciosa* but the beard is darker brown, and the bracts always clear pink with satin-like sheen, and above Betty's Bay it is shot with blue and mauve. The leaves are more round on an erect tree about 2 metres high. It flowers in late winter.

Protea grandiceps Trattinick

This is another protea from the high mountains where, unfortunately, it has been practically exterminated by fires, but it can still be found in the mountains near Elgin and Swellendam. It was first discovered on Devil's Peak and it was also found in the Jonkershoek mountains until it was destroyed there by fire in 1951. Fortunately it responds well to cultivation and is becoming popular in gardens. The shrub is less than $1\frac{1}{2}$ metres in height and has attractive foliage. The flowerheads are terminal and erect and of unusual colour, either peach or coral-red. It flowers in spring.

Protea pulchra Rycroft

Although the bracts are smooth and shiny there is a fine fringe of hair at their tips. Their colours may be found in any shade from apple-green to bright red. It flowers in early spring and the fast-growing new branches soon surpass the older so that they extend beyond the flowerheads, as can be seen in the photograph. This also happens in some other species (e.g. *P. repens*, *P. obtusifolia*). It really belongs to the Cape Flats where it had to make way for crops and vineyards but it is still fairly common in the foothills of the Drakenstein and Hottentots-Holland mountains. The shrub is only about 1 metre in height, and rather spreading. It was formerly known as *P. subpulchella*.

Protea macrocephala Thunberg

Can be seen growing next to the roadside near Humansdorp where it flowers in spring on a slender shrub 3 metres tall.
It is no longer common elsewhere along this coastline. The form growing near Grabouw has a red ring around the protruding flower-tuft (bottom picture).
The colour of the bracts is always apple-green.
It was formerly known as *P. incompta*.

Protea barbigera Meisner

This is the showiest of them all with its bracts edged with powder-puff down. Its natural habitat is also high in the mountains where it is rare, but it is becoming popular in gardens. After *P. cynaroides* it has the second-largest flowerhead measuring up to 20 cm across. (All the other proteas described so far have flowerheads which measure approximately 12 cm in length.)

The tree is only about 1½ metres in height and fairly spreading. Like many other species of Proteaceae it shows slight differences from one locality to another. Some of these local variations or, more correctly, forms are shown on the following pages. *P. barbigera* is commonly known as the woolly-bearded protea, also the Queen Protea, and the form on page 20 is known as the Prag van Tulbagh (Pride of Tulbagh).

Protea barbigera

Cedarberg.

Protea barbigera

Bokkeveld.

Protea barbigera

Du Toit's Kloof.
Possibly now extinct as a result of repeated fires in this area.

Protea barbigera

Franschhoek.

Protea barbigera

Hermanus.

Protea barbigera

Klein Swartberg.
This interesting form was discovered recently high in the mountains near Ladismith.
Formerly it was not believed to exist so far east. The colours
here are either cream or pink, all with dark-brown beard.
Subsequently most of the trees have been destroyed by fire.

SECTION EXSERTAE

The proteas in this section have dainty flowerheads measuring, on average, only about 8 cm in length, but they grow on untidy-looking shrubs which are about 3 metres tall or more, and rather straggly. They are obviously closely related and hybridize easily when grown in proximity. All flower in autumn, except *P. glabra*, which I am not sure should be included in this section.

Protea punctata Meisn.

Is common in the Swartberg and is also found in the Cedarberg. It is best to pick buds when the bracts show the first signs of opening. They will then open in the vase, in contrast to the majority of proteas which do not open further after being picked. The usual colour is pink, but it may also be white.

Protea mundii Klotzsch

Is found along the Tzitzikama coast. I have also come across it in the Hottentots-Holland mountains. The little flowerheads are cup-shaped and there are distinct little knobs to the tips of the styles. The two colours shown do not, of course, occur on the same tree. After they are picked the cups tend to close slightly.

Protea longiflora Lamarck

Has a larger flowerhead than the other proteas in this section and measures about 12 cm across. The long styles are arranged in funnel-shape and are the same colour as the bracts, which may be pink or white. It is not as long-lasting. It is fairly common in the Outeniqua mountains.

Protea lacticolor Salisb.

Comes from the Katberg. It appears to be very closely related to
P. *mundii*, but the tree is sturdier and scarcely more than 3 metres in height whereas
P. *mundii* may reach a height of 6 metres. It also lacks the tiny
knobs on the tips of the styles. The colours may be white or pink.

Protea lanceolata E.Mey. ex Meisn.

Is shown here at its most attractive stage. When the styles open the little flowerhead looks spidery and untidy. It grows in the Mossel Bay area. The colour is always white.

Protea glabra Thunb.

Grown near Citrusdal. It has the distinction of having a scent, which is most unusual for a protea. Note that the bracts are shorter than the flowers, which is typical of proteas from the summer-rainfall area (page 39). It flowers in spring.

SECTION PINIFOLIAE

Proteas with narrow or needle-like leaves are grouped together in this section, but the dividing-line is not clear-cut and the classification is unsatisfactory. Most of them have small, rose-shaped flowerheads which are popularly, but loosely, known as Mountain Roses.

Protea nana (Berg.) Thunb.

The little flowerheads hang face-down like tiny bells, only 3 cm wide, on a free-flowering, compact bush up to 1 metre in height. The popular name 'skaamrosie' (bashful little rose) is very apt. It flowers in winter in the high mountains near Paarl and Worcester.
Linnaeus named it *P. rosacea*.

Protea pityphylla Phillips

Comes from Michell's Pass near Ceres, and is rare. The shrub is sprawling as the long branches bend over to lie near the ground. The terminal flowerheads measure about 8 cm across and hang face-down. It flowers in winter.

Protea witzenbergiana Phillips from the same area is rather similar but the flowerhead is smaller, measuring only about 5 cm across.

Protea canaliculata Haw.

Is found in the Langkloof area but is very rare. The bush is low and sprawling with the branches lying practically on the ground. The narrow grooved leaves, curving upwards from the stem, are about 15 cm in length. The terminal flowerheads are about 9 cm wide and face slightly downwards. The bracts are cream-coloured, shaded pink on the parts exposed to the sun, and have a satin-like sheen.

Protea odorata Thunb.

This little protea has the smallest flowerhead, measuring only 3 cm across. It also has a scent. It was described by Andrews from a plant cultivated in England, but here at the Cape it was believed to have become extinct, until 150 years later when Prof. H. B. Rycroft, after a long search, discovered a few plants near Malmesbury. The shrub is spreading and only about ½ metre in height. It flowers in early autumn.

Protea harmeri Phillips

Seems to prefer the north-facing slopes of the mountains and is found from Hex River to the Swartberg. It flowers from autumn till early spring and the 5 cm wide, dark-red, rose-shaped flowerheads face upwards on a shrub almost 2 metres in height.

Protea scolymocephala (L.) Reichard

Comes from the Peninsula and the Cape Flats and is still fairly common near Kommetjie. The shrub is about 1 metre in height and the greenish-yellow flowerheads measure about 4 cm across. It flowers in early spring.

Protea cedromontana Schlechter

Is found, as the name indicates, in the Cedarberg, but it also grows near Villiersdorp. The little flowerheads measure hardly more than 3 cm across and are rose-shaped and deep red in colour. They face upwards on a tree which may be 3 metres in height.

Protea pendula R.Br.

Although the tree is erect, it is only about 1 metre high and the branches are curved near their ends so that the little flowerheads hang face downwards. They measure about 5 cm across and are rose-shaped. The bracts are reddish-brown in colour and covered on their outsides by silky hairs. It grows near Ceres and is rare. (Photo. by Prof. H. B. Rycroft.)

Protea aristata Phillips

First discovered in 1928 by Stokoe, this protea could not be found again and was believed to be extinct until, 25 years later, it was rediscovered in the Seweweekspoort near Ladismith. The shrub looks like a small pine tree, only about 1 metre high, and the cup-shaped flowerhead, which has ligulate bracts, is about 12 cm long. It flowers in summer. The red flowers are white-tipped.

Protea minor Compton

Has a dainty cup-shaped little flowerhead which measures only about 6 cm in length. The shrub is sprawling and the branches tend to topple over so that the little proteas lie on the ground. The leaves are upright. It comes from the Bredasdorp area and flowers in winter.

Both proteas on this page have rose-shaped flowerheads which measure about 8 cm across, but their leaves are too broad to classify them under the section Pinifoliae. Both come from the Ceres area where they flower in spring. Both are low, spreading shrubs whose flowerheads face slightly downwards,
and both are uncommon.

Protea effusa E.Mey. ex Meisn.

Is better known as *P. marlothii* Phillips, but this is a later name.

Protea recondita Buek ex Meisn.

Protea sulphurea Phillips

Proteas are uncommon in the Karoo, but this lovely and little-known species is found high on the inland-facing slopes of the mountains between Worcester and the Swartberg. The flowerheads are saucer-like in size and shape and hang face downwards on a low spreading shrub, which appears to be very slow-growing and to take several years before coming into flower. The outsides of the smooth overlapping bracts are purple in colour but green at their bases and red along their edges. The effect on the back of the flowerhead is of a remarkable uniform pattern such as one might expect in a circular stained-glass window. It flowers in midwinter when frost is usual and snow not uncommon.

Protea venusta Compton

Is a rare protea which grows in the Swartberg at a height of more than
1 500 metres above sea-level. The shrub is low and spreading, the
branches lying along the ground, but they curve upwards towards their ends so that
the terminal flowerheads face upwards. These are cup-shaped, and only
about 7 cm long. The bracts are ivory-coloured but shaded pink where the edges
have been exposed in the buds. The flower-tuft is pure-white and satin-like. The effect
is one of exquisite daintiness and purity. It has been cultivated lower down but then
appears reluctant to come into flower. The flowering time is in midsummer.

Both proteas on this page are more like those of the summer-rainfall area in that the flowers are longer than the wide-opening, bowl-shaped bracts. Proteas with this type of flowerhead can be grouped together in a section to accommodate the remaining species.

Top left: *Protea aborea* Houttuyn.

The 'waboom' is widespread throughout the mountains but seems to prefer the more sheltered and warmer kloofs of the lower slopes. The tree is about 3 metres high and is sturdy. The leaves have a bluish-green colour and even more attractive than the flower are the new leaves, at the tips of the branches, which are bright red. Near Villiersdorp is a rare form with red flowerheads.

Below: *Protea rupicola* Mund. ex Meisn.

Is rare and grows in the mountains near Jonkershoek where it flowers in early summer.

It seems to be synonymous with *P. dykei* from the Cockscomb mountains near Uitenhage. Proteas are usually at their best before the flowers open, but here, I feel, we have an exception as the long styles are bright-red tipped with yellow, which makes this protea spectacular and most unusual.

PROTEAS OF THE SUMMER-RAINFALL AREA

Protea multibracteata Phillips

The majority of species of this area are rather similar. They have short bracts which open wide and are dominated by the longer flowers. On the whole their colours are not as delicate as those of the south-western Cape, and they are not as striking. *P. caffra* Meisn., *P. rhodantha* Hook, *P. simplex* Phillips, and *P. transvaalensis* Phillips resemble each other closely, but the last two are low-growing, whereas the others are sturdy, gnarled trees about 4 metres high. Also similar, but not so tall are *P. welwitchii* Engl. and *P. gaugedii* Gmelin.

P. rubropilosa Beard, with its short, brown bracts, reminds me of *P. glabra* (page 29) but the flower-tuft is pink.

P. comptonii is white and larger, but otherwise very similar.

right
Protea welwitchii Engl.
was previously called *P. hirta*
(Photo by Mr. H. C. Scholtz.)

PROTEAS WITH MAIN STEM UNDERGROUND

Protea tenuifolia R.Br.

The narrow cylindrical leaves are about 15 cm long and the little bowl-shaped flowerheads, which are about 5 cm wide, are situated at ground-level. This specimen was found near Heidelberg. It flowers in early spring, and is rare.

Protea scabra R.Br.

Is fairly common in the mountains of the south-western Cape. The narrow leaves are about 20 cm in length and have a rough feel to them, and partly hide the little flowerhead, which is situated at ground-level in their centre. It is about 5 cm long. It flowers in late winter.

Protea aspera Phillips

Is very similar, but the flowerhead is twice as large and has a longer stem, and the perianths have a more woolly appearance. It is confined to the Bredasdorp area.

Protea lorea R.Br.

When first I came upon this remarkable protea I felt that my leg was being pulled. It looked as if someone had picked a protea and stuck it into a clump of thatching-grass. The cup-shaped flowerhead is as much as 12 cm in length and the cylindrical leaves are about 30 cm long. It grows on the lower slopes of the mountains near Somerset West, but this photograph was taken in the mountains near Heidelberg. It is little known, probably because it flowers in midsummer when it is too hot for climbing to be popular.

Protea acaulis (L.) Reichard

The branches run just under the surface of the ground with only their tips emerging where the leaves are situated, so that the impression is often of two or three separate plants growing close together. The flowerheads are situated at ground-level and have practically no stems. They measure about 5 cm across. The usual colour of the bracts is yellowish-green with pink edges and the red colour is uncommon. It flowers in early spring. It has a fairly wide distribution in the mountains and is common near Franschhoek. The broad leaves are usually about 12 cm in length but their size and shape vary from one locality to another.

When I came across the specimen, shown below, in the mountains of the Langkloof I thought it was a new species, and was surprised to discover that it was *P. acaulis* var. *cockscombensis*. The Cockscomb Peak is near Uitenhage.

Protea glaucophylla Salisb.

This is a rare protea which grows in moist patches in the Koue Bokkeveld and Cedarberg mountains at a height between 900 and 1 200 metres. The leaves are about 20 cm long, dull greyish-blue in colour with the margins and midribs often flushed with pink. The flowerheads are about 4 cm wide. It flowers in late spring. (Photo. by Mr. H. C. Scholtz)

Protea amplexicaulis R.Br.

The branches lie on the ground and have neat sessile leaves edged with red. The new leaves at the ends of the branches are deep red so that the plant is attractive in itself. The small flowerheads measure about 7 cm across, and are hidden under the branches near the main stem and face downwards practically on to the ground. The bracts are dark brown and velvet-like on their outsides but smooth and cream-coloured on their insides.
It flowers in winter in the mountains near Du Toit's Kloof.
Protea humiflora Andrews, also has furry bracts but they are red. The flowerheads are also situated near the main stem but the leaves are narrow, almost needle-like. It grows in sandy soil in the Caledon area.

Protea acerosa R.Br.

The small flowerheads are clustered near the main stem close to the ground. They measure about 4 cm across and also have furry brown bracts. The plant is about 30 cm high and has needle-like leaves.
It is found near Betty's Bay where it flowers in late winter.

P. caespitosa. Andr. (right)
This rare protea from the high mountains above Jonkershoek differs from other proteas in that the red bracts are surrounded by brown, roundish leaves without stalks whereas the normal leaves are long with long leaf-stalks. The flowerhead is about 5 cm in width. It flowers in early summer. It is also known as *Protea turbiniflora* R.Br. (Photo. by Mr. S. W. Chater)

Protea decurrens Phillips (top left)
The little flowerheads, about 4 cm wide, are hidden deep inside the plant in clusters near the main stem and are of a pale, almost translucent, salmon-pink colour. It flowers in late winter.

Protea scolopendrium R.Br.
Is found in the mountains of the south-western Cape but is not common. It flowers in early spring. The long leaves measure about 20 cm and the stiff flowerheads are about 8 cm wide. A dwarf variety grows high in the Swartberg.

Protea cordata Thunb.

The little flowerheads, only about 3 cm wide, are clustered around the main stem at ground-level. More attractive are the large, heart-shaped leaves on long, red leaf-stalks. It grows in the mountains near Bot River and flowers in late winter.

Protea restionifolia (Salisb. ex Knight) Rycroft

This is a rare protea which grows on a 'Karoo-koppie' near Worcester. The leaves look like stiff, hairy grass and are about 20 cm long. The little proteas are situated at ground-level and measure about 6 cm across. They open wide in spring.

Another plant which looks like a clump of grass is *P. tugwelliae* from high in the Swartberg. The leaves are finer and more grass-like and the little flowerheads, only about 3 cm wide, are reddish-brown in colour, and situated at ground-level.
P. montana is similar but a little larger.

Protea cryophila Bolus

This is a very rare protea found in only one or two spots high in the Cedarberg near Ceres. The large leaves may be as long as 60 cm and are folded along their length. Side-branches from the underground main stem are usually also just under the surface of the ground so that the plant may spread over an area about 3 metres wide. The flowerhead is about 12 cm in length and the outsides of the bracts are white and woolly but they are smooth and deep red on their insides. Until fairly recently it was generally believed that the flowering time was in December and that the bracts do not open much more than shown on the photograph, but it has since been found that in February they open much wider. It is known as the snow flower.
(Photo by Rev. R. Andrag, by kind permission of Mrs. M. Vogts.)

Protea convexa Phillips

This is another very rare protea which is found in the mountains at Matjiesfontein, near Laingsburg. The leaves are about 16 cm long and have a light coating of powder so as to resemble young cauliflower leaves. They have attractive bright-pink veins. Side-branches also spread underground so that the plant may cover an area about 2 metres wide. The flowerhead measures about 12 cm across and the bracts are smooth and pink on their outsides (the same colour as the veins in the leaves), and on their insides they are coloured bright-yellow. When the styles open they are red. I brought this specimen home and it lasted in water for more than a month. The flowering time is in winter.

The Genus Orothamnus Pappe

Orothamnus zeyheri Pappe

The marsh rose is the only species in this genus. It is extremely rare and in danger of becoming extinct as there are a few plants only in almost inaccessible spots high in the mountains above Hermanus and Betty's Bay, and attempts to cultivate it have not been successful. The plant grows in marshy soil and consists of one or more erect branches as much as 3 metres tall. The flowerheads are of exquisite shape and beauty. Although the pomegranate-red bracts have a covering of fine hairs on their outsides they are, at the same time, smooth and shiny and have a delicate, waxlike translucency. They overlap and are curled back at their edges in the fashion of a rose. It flowers in early summer.
(Photo by Mr. S. W. Chater.)

The Genus Leucospermum R. Brown

Commonly known as 'pincushions', the leucospermums lack the large, coloured bracts of the proteas, but they provide a wonderful display of colour in spring when their flowering period lasts for four months or longer. The shrubs are usually neatly rounded in shape and free-flowering, bearing a terminal flowerhead to each branch. When not in flower they can be differentiated from other Proteaceae in that the leaves are 'toothed', having small indentations with raised edges, often red in colour, at or near their tips. There are about 40 species of which the vast majority occur in the south-western Cape.

Leucospermum spathulatum R.Br.

This attractive species with flame-coloured flowerheads grows in the Cedarberg, where it is rare. It differs from most species in this genus in that the shrub is sprawling with the branches close to the ground and curving upwards towards their tips where the flowerheads are situated.

Leucospermum cuneiforme (Burm.f.) Rourke

Was formerly known as *L. attenuatum*. At Swellendam the flowerheads are yellow with light-orange styles on a well-rounded shrub which reaches a height of about 1½ metres. On the lower slopes of the Swartberg and near Humansdorp and Port Elizabeth the flowerheads are always yellow, and a little smaller on a lower shrub, whereas at Knysna the colours change from yellow through orange to red as the flowers mature, and at one spot near Knysna it is always red.

The most colourful form is shown below and is found at Albertinia and Mossel Bay, but this is now regarded as a separate species and will probably be given a different name. The tree is sturdy, reaching a height of about 2 metres or more. It has a well-defined main stem and a deeper root-stock so that it is less easily destroyed by fire.

Leucospermum vestitum (Lam.) Rourke

Was formerly known as *L. incisum*. The rolled-back perianths around the central cone are red or purplish in colour usually, and the 'pins' or styles are yellow, but colour is of little import in determining a species. The leaves are more deeply toothed and the flowerheads are more flat-topped. It grows in the Citrusdal area.

Leucospermum glabrum Phillips

Grows near George where it is now rare in Nature. The tree is more sturdy and more than 1 metre in height. As its name suggests the leaves are smooth. The flowerheads always have the same colours as in the illustration.

Leucospermum conocarpodendron (L.) Buek

The tree is also sturdy and may be 3 metres high. It is found in the Peninsula near the sea and further east. The flowerheads are always of a bright golden-yellow colour. The name used to be *L. conocarpum*.

Leucospermum cordifolium (Salisb. ex Knight) Fourcade

Is well known under its old name *L. nutans*. It is most common in the Caledon area. The shrub is about 1 metre in height and the flowerheads are about 10 cm wide.

Several species are practically indistinguishable from it, the only difference being in the leaves or the habit of the shrub. *L. vestitum* is shown on the opposite page.

L. patersonii Phillips is taller in habit and *L. bolusii* Phillips is no longer recognized as being a different species. The most common colour of the flowerheads is orange.

Leucospermum tottum (L.) R.Br.

This is one of my favourites. The neatly rounded shrub is about 1 metre high and practically covered with flowers. The perianths do not roll back all the way on to the 'hub' and this gives the flowerheads a lacy and softer appearance.
It grows in the mountains near Du Toit's Kloof but is not common.

The younger flowerheads are pale salmon-pink in colour and become flat-topped when the styles open. These now become yellow, and the perianths red or even purple in colour.

Leucospermum lineare R.Br.

Differs in that the leaves are narrow and almost needle-like. It is also found in the mountains near Paarl and Franschhoek where the commonest colour of the flowerheads is an unattractive greenish-yellow, but it may be deep red at the other extreme.

Leucospermum reflexum Buek ex Meisn.

The flowerheads remind me of a volley of sky-rockets being sent heavenwards. As the styles separate from the perianths they both fold over backwards. The leaves are small and silvery but the tree may become as tall as 4 metres. It comes from the Cedarberg and may flower for as long as six months. It is commonly known as perdekop.

Leucospermum catherinae Compton

Here we see the styles unfolding in an anti-clockwise direction as if by circular rotation so that the open flowerhead resembles a spinning Catherine-wheel.
At a later stage they turn upwards and then remind me of a Roman candle. Perhaps the pincushions could popularly, and just as descriptively, be known as 'floral-fireworks'.
It grows near Citrusdal. The styles are straw-coloured tipped with pale lilac stigmata.

Leucospermum grandiflorum R.Br.

Grows on the lower slopes of the mountains near Somerset West and appears to be closely related to *L. catherinae*. It is more colourful, several shades or colours being present at the same time, representing different stages of maturity, but the general effect on the shrub is rather stiff and untidy.

Leucospermum gueinzii Meisn.

This rare and fascinating species appears to be closely related to the above and its location in the Jonkershoek mountains is only a few miles further north.

LEUCOSPERMUMS WITH SMALL FLOWERHEADS

On average these measure only about 3 cm across.

Leucospermum truncatulum (Salisb. ex Knight) Rourke

Was formerly known as *L. buxifolium*. It is very similar to *L. calligerum* (opposite) but the shrub is more erect in habit and the leaves slightly broader
and more closely set to the stem so that they overlap.
I find it more colourful. It grows in the mountains near Bot River.

Leucospermum muirii Phillips

Can be seen growing next to *L. cuneiforme* at Albertinia (page 49)
and, at first sight, it appears to be a dwarf form of the latter.
It goes through the same colour changes as the little flowerheads mature. The shrub is, however, quite large and may reach a height of more than 1 metre.

Leucospermum calligerum (Salisb. ex Knight) Rourke

Was formerly called *L. puberum*. It has a wide distribution throughout the south-western Cape and I've even come across it high in the Swartberg where snow is not uncommon. It has a long flowering period, but the colours are never very bright.

Below, right:
Leucospermum oleaefolium (Berg.) R.Br.

Was formerly known as *L. crinitum* but the latter is now regarded as a form and not a separate species. It is most common in the region of Betty's Bay. The shrub may reach a height of about 1 metre and is most colourful as the flowerheads, arranged in groups of three or four, pass through all warm colours and the various stages can be seen in a single group. It is in flower for many months.

Below:
Leucospermum mundii Meisn.

From the mountains near Riversdale, looks very similar but the flowerheads are a little larger. It is rare.
(Photo. by Mr. S. W. Chater)

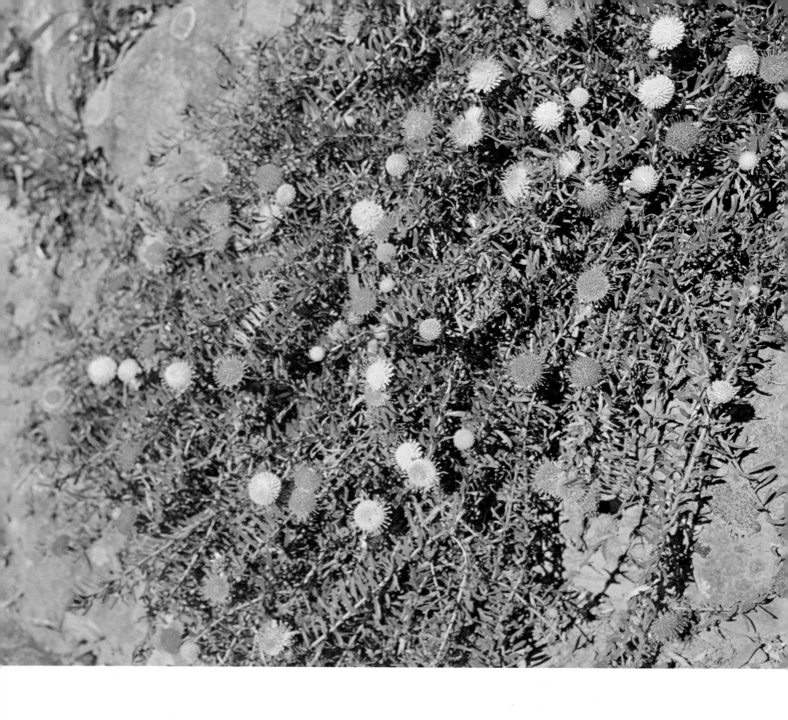

Leucospermum prostratum (Thunb.) Stapf.

Is a creeper with long trailing branches. It grows in sandy, acid soil in the Bredasdorp area. The little flowerheads are only about 2 cm wide. It flowers practically throughout the year.

Leucospermum bolusii Gandoger

Was formerly called *L. album*. The neatly rounded shrub is about 1 metre high and covered with the little white flowerheads, but it fails to attract much attention as, from a distance, it looks rather dull.
It grows in the Gordon's Bay area.

Below, right:
Leucospermum rodolentum (Salisb. ex Knight) Rourke

This is a tree about 4 metres high which has small silvery-grey leaves. It grows in sandy soil near Villiersdorp but is not common. The little flowerheads are always canary-yellow in colour. It was formerly known as *L. candicans*.

Below, left:
Leucospermum hypophyllocarpodendron (L.) Druce

The little flowerhead looks similar but, in contrast, the small shrub is trailing.
It grows in sand near the shore at Blouberg and usually only the erect leaves and little flowerheads protrude above the sand.
It was formerly known as *P. hypophyllum*.
(Photo by Mr. H. C. Scholtz.)

The Genus Mimetes Salisb.

In this genus the flowers are situated in the axils of the upper leaves of the erect branches. These upper leaves are sometimes brightly coloured.
There are 11 species, all from the south-western Cape.
Attempts to cultivate the mimetes have not met with much success. They have very tiny seeds and it is difficult to get them to germinate. In the case of *M. cucullata* there has been more success by treating cuttings with rooting hormone but this is, in fact, the only species I have seen successfully cultivated in any garden.

Mimetes cucullata (L.) R.Br.

Was formerly known as *M. lyrigera*. It is quite common in the Betty's Bay area and vicinity but is also found in the mountains and is fairly widely distributed. The shrub is about 1 metre in height but *M. hartogii* R.Br. which was formerly thought to be a variety, grows to a height of about 3 metres. It is popularly, but rather unpoetically, known as 'rooistompies'.

Mimetes hirta Knight

Is rare, and grows in marshy ground at Betty's Bay. Here the upper leaves remain green and the flowers are ensheathed in largish bracts which provide the colour.

Mimetes pauciflora R.Br.

Grows in the mountains near George but is rare. Here again the colour is provided by the long bracts which encase the flowers. These bracts are bright orange, tinted red on the sides facing the sun.

Mimetes argentea Knight (right)

Is rare and grows in the mountains near Villiersdorp. The tree is about 2 metres high and has silvery leaves. The upper leaves are shaded red which deepens towards their axils where the flowers are situated. When these open the red styles, with yellow tips, lie along the under surfaces of the leaves and the perianths are raspberry-coloured twirly tufts lying on the base of each leaf. The effect is remarkable. It looks as if tiny, red electric globes are hidden amongst the upper leaves, their light being reflected by the silvery leaves.

Mimetes hottentotica Phillips (below, left)

Is very rare and grows high in the Hottentots-Holland mountains. The tree also has silvery leaves but the upper leaves do not become coloured. It is about 2 metres high. The red styles are tipped with black stigmata and the periantheal tufts are yellow.

Mimetes capitulata R.Br. (below, right)

From the same area, is the rarest of all, and there are only a few plants. These are less than 1 metre in height and the leaves also have a silvery sheen. The flowers are encased in brightly coloured bracts. (The 3 photographs on this page are by Mr. S. W. Chater.)

The Genus Serruria Salisb.

The serrurias have finely divided leaves and the flowerheads are
placed in groups between the leaves at the ends of the branches to form clusters.
They flower in early spring. There are about 50 species.

Serruria florida Knight

The blushing-bride is the most outstanding species in this genus. Unlike the others it has comparatively large bracts so that the little flowerheads look like small proteas measuring only about 4 cm across. They have a delicate, paper-like translucency and are white, blushed with rose. The slender plant reaches a height of about 1½ metres. It is found only in the Franschhoek mountains where it is very rare and for as long as a century it was thought to be extinct, but it is now popular in gardens after having been cultivated and preserved at the National Botanic Gardens at Kirstenbosch. The young plants may start to flower after only nine months.

Serruria aemula R.Br. (formerly *S. rosea*)

Comes from Wemmershoek where it is now practically extinct due to the cultivation of pine trees and infestation with Hakea. The shrub is compact and only about ½ metre in height. The bracts are smaller and of deeper colour and the little flowerheads more closely clustered than those of *S. florida*.

Serruria barbigera Knight

The bracts are insignificant and the hairy flowers dominate. The shrub is erect and only about ½ metre in height. It grows in the mountains in the Caledon area.

Serruria elongata R.Br.

The cluster of flowerheads is supported by a long stem which measures about 20 cm in length, and the tiny bracts are hardly visible. It grows in the mountains of the Caledon area.

Serruria pedunculata (Lam.) R.Br.

Was formerly called *S. artemesiaefolia*. It is found in the Clanwilliam area and the shrub is about 1 metre in height. The little flowerheads are not very colourful and are spread across the top of the plant on short stems. They are only about 3 cm wide.

Serruria adscendens R.Br.

The plant is small and rather spreading and the tiny flowers, in lilac-coloured clusters, have a scent. It is fairly common in the Betty's Bay area.

Serruria burmanii R.Br.

The lilac-coloured clusters are small and composed of very tiny flowerheads and may be quite dense on a shrub which may be near to 1 metre in height. In the majority of serrurias the bracts are not visible on the open flowerheads.

The Genus Paranomus Salisb.

In this genus the individual flowers are arranged in groups of four, each group being surrounded by four bracts, two inner and two outer, to form a small flowerhead. These are clustered on spikes at the ends of the branches. As the name indicates, some of the species have two kinds of leaves (dimorphic). There are 14 species, found from Clanwilliam to Uitenhage. They flower in late winter.

Paranomus sceptrum—gustavianum (Sparm.) Hylander

The leaves on the lower part of the plant are finely divided but they are broad on the stems leading to the terminal flower-spikes. These are about 8 cm in length and always white in colour. It grows in the mountains in the Caledon area and is rare.

Paranomus reflexus (Phillips & Hutchinson) N.E.Br.

This is a handsome shrub, about 2 metres high, which grows in the Port Elizabeth–Humansdorp area. Like *P. sceptrum—gustavianum* the lower leaves are finely divided and the leaves near the flower-spikes are broad. The flowers are greenish-yellow in colour and are reflexed on the spike.

Paranomus spicatus (Berg.) R.Br.

Has only one kind of leaf. This is finely divided so that the plant looks like a serruria. The spikes are coloured mauve and pink and are only about 8 cm in length. It grows in the Caledon and Gordon's Bay areas.

P. bracteolaris Salisb. ex Knight

Is a larger species of similar appearance which grows in the Citrusdal area. It was formerly known as *P. crithmifolius*.

The Genus Aulax Berg

This genus is dioecious—the male and female flowers being borne on separate plants. The female flowers are surrounded by bracts which makes them resemble small proteas, whereas the male flowers are borne in fluffy clusters on short spikes. There are only three species. They flower in summer.

Aulax cneorifolia Knight

Is relatively common in the Hermanus area. The tree is about 2 metres in height. It has fairly narrow leaves and the new leaves have a reddish colour. The male flowers are shown on the left and the female flowerheads, which are about 3½ cm across, are shown below.

Aulax pinifolia Berg

Has needle-shaped leaves. It is rare, but one or two good patches can still be found near Knysna (left).
The bracts of the female look feathery but they are as hard and stiff as the bristles on a toothbrush. They are retained, after the flowering stage, to enclose the seeds and so form a seed-bearing 'cone' which may become bright red in colour, as can be seen in the picture below, and may be mistaken for a bud. The top leaves surrounding the flowers also have a tendency to become bright red in colour. The male flowers are shown below on the left. There are usually more spikes to a cluster.

Aulax pallasia Stapf.

From the Worcester area. Has leaves which are not quite as thin, but not as broad as those of *A. cneorifolia*.

The Genus Brabeium Linnaeus

Brabeium stellatifolium (wild almond)

This is the only species in this genus. It is a handsome tree about 5 metres in height which grows along the streams in the south-western Cape. The clusters of white flowers have a scent and are borne on spikes which radiate from a whorl of leaves which are about 15 cm long. The young leaves are velvety and dusty-gold in colour.
This tree was chosen by Jan van Riebeeck when he planted a hedge in 1660 to demarcate the boundaries of his settlement. Parts of this hedge are still to be seen today and the trees still look vigorous. One part is preserved at Kirstenbosch.

In all the South African Proteaceae the seed is a tiny nut, but in the Wild Almond this is a large drupe resembling an almond. It is poisonous but becomes edible after soaking, and was used by the early colonists as a substitute for coffee.

The Genus Leucadendron

This is the second largest genus with close to 100 species. Like the Genus Aulax it is dioecious and the male and female flowers are borne on separate plants, which may look quite different. The tiny female flowers are situated on a seed-forming 'cone' (tolbosse), and the male flowers are small, fluffy 'pompons'. In most species the terminal leaves surrounding the flowers become bright yellow during the flowering period, and in several species this changes to orange and red so that the bush presents a different appearance at different times of the year. This, added to the fact that there may be slight differences from one locality to another, has resulted in a certain amount of confusion. This genus is still not well understood and many species have been incorrectly named. The only valid name is the one given with the first published description. To Mr. Ion Williams has been allotted the formidable task of studying and revising this genus. The vast majority of the species are found wild only in the south-western Cape. Most of them flower in late winter.

Leucadendron microcephalum Gandoger & Schinz

Many species colour the veld with patches of bright yellow in winter but this one can be differentiated by the fact that the flowers are surrounded by a little ring of brown bracts. Mr. Ion Williams is seen here at work in the Hottentots-Holland mountains. This species was formerly known as *L. stokoei*.

Leucadendron gandogerii Schinz ex Gandoger

Was formerly known as *L. guthrieae*. It is fairly common in the Caledon area. The shrub is usually more slender and about 2 metres in height. The male flowers are shown on the left. Rather similar but with a larger inflorescence is *L. saxatile* from the Cape Peninsula. This name has now been changed to *L. strobilinum* (L.) Druce.

Here the terminal leaves have returned to green after the flowering period and the male flowers blacken before dropping off. The new shoots grow out from just below the flowers and above the terminal leaves.

This is the female. The leaves are returning to green after the flowering period and the seed-bearing 'cone' is now developing.

Leucadendron laureolum (Lam.) Fourcade

Was formerly known as *L. decorum*. The terminal leaves enclose and hide the flowers so that you have to open them before you can tell whether the plant is male or female. They become bright yellow tinged light orange, and this colour appears to be richer in the male. It is found in the mountains near Bot River.

It is fairly common in the mountains near Houhoek.

Leucadendron adscendens R.Br.

This name will have to be changed to *L. salignum*. It is so common in the south-western Cape that it is generally regarded as a weed (geelbos). Although it is generally low-growing it might reach a height of nearly 2 metres. The male flowers are shown on the left. The terminal leaves are always narrower in the male than in the female, where they broaden towards the base, but in the form from Knysna, seen here, they are even narrower than usual.

The terminal leaves hide the female 'cones' in the flowering stage. They are usually bright yellow, and the red edges, shown here, are unusual.

The deep red variety is less common but can be found growing in proximity to the common yellow variety. It is at its best in the Langkloof where it is known as 'rooibos' (below).

Leucadendron xanthoconus (O. Kuntze) K. Schum.

Was formerly known as *L. salignum*. It is also common, especially in the west, and is also popularly known as 'geelbos'. The tree is about 2 metres in height.
The male is shown above and the female is shown below.

Leucadendron tinctum Williams

Well known under its former name, *L. grandiflorum*, I find this one of our most fascinating leucadendrons. The shrub is about 1 metre in height and the inflorescence measures about 10 cm across. It is fairly widely distributed throughout the mountains of the south-western Cape but is not very common anywhere.

I use this as an example of how confusing some of the members of this genus may be and, where there are variations from one locality to another plus the fact that the same shrub presents an inflorescence which looks different at different times of the year, it is not surprising that so many of the leucadendrons have had their names changed when it was found that they have been described and named before. Local variations, or forms, of *L. tinctum* are shown on this page. The photographs on the opposite page show various stages at different times of the year, again of various forms.

Top: *Hermanus*
Centre: *Langkloof*
Bottom: *Port Elizabeth*

February (Swartberg)

May (Hermanus)

July (Port Elizabeth)

November (Hermanus)

Leucadendron sessile R.Br.

Was formerly known as *L. venosum*. It is even more colourful than *L. tinctum* and three stages are shown here. The male flowerhead is quite a large 'pompon' which is surrounded by a ring of black bracts. The female 'cone' is red and shiny.

Leucadendron daphnoides (Thunb.) Meisn.

Is another lovely species whose terminal leaves change colour through all the autumn shades in the flowering period. They hug the flowerhead before curving outwards. It is seen here in the mountains near Du Toit's Kloof.

Leucadendron humifusum
Phillips & Hutchinson

Was formerly known as
L. cordatum. This is, I think, the
loveliest of all leucadendrons.
The flowers are rather similar to
those of *L. tinctum* but the shrub is
low with the branches bending
over low to the ground so that
the flowers hang facing
downwards. Not only the terminal
leaves surrounding the flowers,
but also the distal ones near the
end of the branch, become
brightly coloured in the flowering
period, changing from yellow to
orange, to bright red, and then
even to purple, and all these
colours may be seen on the same
shrub at the same time. These
coloured leaves are smooth
and look delicate but they are,
in fact, hard and stiff.
The flowers are also exceptional in
having a scent.
It is very rare and grows near
Montagu where it is known as
'Bergkatjiepiering'. Recently it
was also found in the
Swartberg. It appears to be very
slow-growing.

Leucadendron comosum
(Thunb.) R.Br.

Was formerly known as *L. aemulum*. It grows in the mountains of the Langkloof and it can be seen in the Swartberg Pass, near the turn-off to Gamkaskloof, where it flowers in December. It has needle-like leaves but these broaden out towards the end of the branch so that the terminal leaves surrounding the female 'cone' are comparatively wide. They become light yellow during the flowering period and are arranged in a neat bowl-shape to simulate overlapping petals. The seed-bearing 'cone' eventually becomes as large as that of a pine tree. The inflorescence on the female plant is quite large and measures about 10 cm across. The tree is erect and about 2 metres high.

The male inflorescence is much smaller, measuring only about 3 cm across. The terminal leaves do not broaden out as much and the inflorescences resemble clusters of small daisy-like flowers. As in many other species, the male is more profuse in flowering.

Leucadendron eucalyptifolium Buek ex Meisn.

The shrubs grow to a height of about 3 metres. The flowering is quite profuse so that it is covered with the small 4 cm wide inflorescences which have a 'winged' appearance since the flowers are situated in a little cup of short leaves from which longer leaves also radiate. Not only do these terminal leaves become brightly coloured but the whole tree becomes yellow in the flowering season, and since it can form quite dense vegetation it colours the veld with patches of bright yellow near Knysna, where it is common.

Below, right:
Leucadendron salicifolium (Salisb.) Williams

Was formerly known as *L. strictum*. The inflorescences are similar in structure, but very much smaller measuring only about 2 cm across, so that the erect tree has a delicate feathery appearance when it becomes bright yellow in the flowering period. It reaches a height of more than 2 metres and is quite common from Bot River to Bain's Kloof.

Leucadendron platyspermum R.Br.

Is quite common in the Bot River mountains. Although a poor specimen is shown here it serves its purpose and makes a description of the 8 cm wide inflorescence unnecessary and it is easy to identify. The seed 'cone' eventually becomes about 6 cm long and is extremely hard.

Leucadendron nervosum Phillips & Hutch.

Is rare and grows in the Worcester area. The tree is about 2 metres in height. The inflorescences are about 4 cm wide and are unusual in that the leaves have a covering of fine woolly hairs. The colours of green, yellow, and brown also make an unusual and attractive combination.

Leucadendron uliginosum R.Br.

Grows in the Outeniqua mountains where it is relatively common. The leaves are small and have an attractive silvery-grey colour. The shrub is about 2 metres in height. The male inflorescences look like yellow everlastings and are borne in clusters not only at the end of the long, thin stems, but also at one or two points lower down the stems. The female inflorescences are quite a bit larger, measuring about 6 cm across, and have a daisy-like appearance. Both are shown here in the same photograph. It flowers in October whereas most leucadendrons are at their best in July.

Leucadendron macowanii Phillips

Is very rare and is found growing wild only at Smitswinkel Bay. The erect tree grows to a height of about 4 metres. It has handsome, smooth, dark-green foliage and somewhat resembles a Black Wattle. The terminal leaves do not change colour in the flowering period. Both male and female flowers are shown here together, but they do not, of course, occur on the same tree.

Leucadendron chamelaea (Lam.) Williams

This pretty little species is found growing in moist sites in the Ceres–Worcester area.
The male inflorescence, shown on the left, is less than 2 cm in width. It was formerly known as *L. decurrens*.

Leucadendron spissifolium (Salisb. ex Knight) Williams

Descriptive of its smooth leaves is its former name, *L. glabrum*. It is not uncommon in the Caledon area. The male inflorescence is shown on the right. It is small and measures only about 4 cm across.
The dainty little female, shown below on the left, is a rare form which grows in the Plattekloof in the mountains near Heidelberg. It is here only that one finds the bright tints edging the whorl of terminal leaves, and it seems hard to believe that these are, in fact, leaves and not petals and that they will become green again towards the end of the flowering period. (Photo by Mr. S. W. Chater.)

Leucadendron elimense Phillips

Grows in the Bredasdorp area. Note how the terminal leaves surrounding the flowers are much larger to give the inflorescence a daisy-like appearance. They become bright yellow in the flowering period. The male is shown on the left. The female, at a later stage, is shown below.
In several species the terminal leaves take on these red tints before they return to green.

Leucadendron modestum Williams

This is a low, neatly rounded shrub, only about 20 cm high, which is fairly common in the Hermanus area. In the flowering period it is covered with small, daisy-like inflorescences which are only about 3 cm in width.

Leucadendron discolor Buek ex Phillips & Hutchinson

Grows near Piketberg and is one of the most attractive species. In the male (above) the 'pompon' is bright orange in colour shading to yellow near its base. It is surrounded by the larger terminal leaves which become bright-yellow and are so neatly arranged that it is hard to believe that they are not petals. This shape is even more attractive in the female (below) where, however, the bright 'pompon' is replaced by a greenish seed-forming 'cone'.
The shrub grows to a height of about 2 metres, and the inflorescence measures about 7 cm.

All three species on this page are from the Clanwilliam area.

Leucadendron procerum (Salisb. ex Knight) Williams

The inflorescence is very similar to that of *L. discolor* but it is smaller in size.

Leucadendron loranthifolium (Salisb. ex Knight) Williams

The female is shown here. The terminal leaves do not change colour but remain an attractive bluish-green colour.

Leucadendron pearsonii Phillips

Was formerly known as *L. glaucescens*. Although the terminal leaves remain greyish-green and do not change colour the whorl gives the inflorescence an attractive shape.

Leucadendron plumosum R.Br.

Has a widespread distribution in the mountains of the south-western Cape. The male and female are shown here. They look quite different. The erect tree is about 2 metres high and, in the male, is covered with tiny 'pompons'. The female 'cone' is coloured brown and silver and the stigmata protrude at the tip of the cone. (The leucadendrons appear to be wind- rather than insect-pollinated.)

The form on the right, which has purple and red 'cones', is uncommon.

Leucadendron dregei
E.Mey ex Meisn.

Is rare and may be found near the top of the Swartberg. It is at its most attractive stage in February when the female 'cones' turn bright coral-red in colour, and resemble tiny rose-buds only about 4 cm long. Although the red colour is retained for several months this very bright phase is brief and lasts for
only about a fortnight.

The shrub is compact and neatly rounded in shape, and is usually less than 1 metre in height.

Leucadendron dregei

The male (above) and the female (below) flowers are shown on this page. They are small and not particularly attractive. The female inflorescence measures only about 4 cm across and has a woolly appearance.
The flowering period is in December.

Leucadendron argenteum R.Br.

Young 'Silver trees' are seen here growing in the National Botanic Gardens at Kirstenbosch which happens to be their natural habitat, as this well-known leucadendron is found growing wild only on the slopes of Table Mountain and nowhere else. They reach a height of about 7 metres. The leaves have a very fine coating of silvery hairs and glitter brightly in the sunlight.
The male flowers are shown above and, below, the 'cone' of the female. It is silvery and slightly larger than a golf-ball, and is spherical in shape.

Leucadendron album (Thunb.) Fourcade

Has narrow, silvery leaves and grows high in the Swartberg and Outeniqua mountains. The 'cone' is peach-coloured and looks like the bud of a protea whose bracts have a satin-like sheen. In fact, when Thunberg found it in the Swartberg he mistook it for a protea and named it *Protea alba*. When his mistake was discovered the name was changed to *L. aurantiacum*, but the valid name is the one given above.
The shrub is erect and about 1½ metres in height. The male flowers are small, dull-yellowish 'pompons', and not very attractive. (Thunberg grouped all 79 species in his collection as belonging to the genus protea, even though they represented eleven genera.)

Leucadendron conicum (Lam.) Williams

Grows in the Outeniqua mountains where it flowers in spring. The inflorescence is small and measures only about 3 cm. The male is shown here and it will look more attractive when the flowers open and the 'pompons' are yellow against the background of small, red bracts.

Leucadendron galpinii Phillips

Can be seen growing next to the roadside near Albertinia. The cone of the female, shown here, measures about 3 cm across and is silvery in colour. The male flowers are clusters of tiny yellowish 'pompons' which are not attractive.

Leucadendron tortum (Thunb.) R.Br.

Is fairly common in the Hermanus area and looks like a dwarf variety of the above. Both the male and the female flowers are shown here.

Leucadendron pubescens R.Br.

Comes from the Cedarberg and is a handsome tree growing to a height of about 3 metres. Although the flowers are not striking I find them interesting because the bracts are fairly prominent and coloured red so that the male flowerheads resemble tiny proteas.
One female and several male flowerheads are shown in the photograph above, but they do not, of course, occur on the same tree. The seed-bearing 'cone', at a later stage, is shown below.

The Genera *Spatalla* Salisb., *Spatallopsis* Phillips, *Sorocephalus* R.Br., and *Diastella* Knight

The species in these genera are, practically without exception, small low-growing shrubs with tiny flowerheads which may be either white, pink or mauve in colour and which are not particularly striking.

SPATALLOPSIS is now regarded as synonymous with Spatalla. The 5 species, which are found from Clanwilliam to Caledon, will no longer be classed as a separate genus and will fall under the genus Spatalla.

The genera on this page have their flowers arranged on short spikes.

SPATALLA with 20 species distributed over the south-western Cape as far as George,

and

SOROCEPHALUS with 11 species between Tulbagh and Swellendam, have their flowers arranged on short spikes.

Top: *Spatalla caudata* (Thunb.) R.Br.
Bottom left: *Sorocephalus imbricatus* (Thunb.) R.Br. (Photo. by Mr. John Rourke)
Bottom right: *Spatalla setacea* (R.Br.) Rourke. (Formerly *Spatallopsis begleyi* Phillips)

DIASTELLA with 5 species between Tulbagh and Caledon, has small, proteaceous flowerheads.

Right:
Diastella serpyllifolia Knight.

The Genus Faurea Harvey

The species in this genus are mostly large trees whose flowers are borne on spikes. They are found from Knysna to the Transvaal, and further north into Africa. There is one species in Madagascar.
There are 5 species in South Africa of which the best known is the 'Boekenhout'—*Faurea saligna* (below). *Faurea macnaughtonii* Phillips is rare and is found in the Knysna forest at Gouna only. It is a tall, slow-growing tree which may reach a height of 20 metres. It is known there as 'Terblanshout'.

Left:
Faurea saligna Harvey.

97

Hybrids

Hybrids are not common in Nature but occur frequently in gardens. This is probably because, where different species do grow in proximity in Nature, they do not usually flower at the same time.

above:
This hybrid obviously occurred in a garden as the two parents are not from the same locality. They are *L. catherinae* and *L. reflexum*.

below:
I came across these two hybrids in the Swartberg where they were growing about 20 kilometres apart. The parents were, in both cases, *P. punctata* and *P. venusta*. The former starts to flower here at the end of the latter's flowering period.
When the same two species are crossed the resultant hybrids will usually not be identical and, should one wish to retain the characteristics, propagation must be done by rooting cuttings from the hybrid.

The hybrid above is probably *P. compacta* crossed with *P. barbigera*.

This is probably *P. longiflora* crossed with *P. barbigera*.

This appears to be a hybrid between *Leucadendron sessile* and *L. daphnoides*.

In the Mountains

The Hottentots-Holland mountains slope steeply down to the sea. During the hot and dry summer months the peaks often become blanketed by mist which forms when the south-easter brings cool air from the sea. Several rare and beautiful species of Proteaceae grow here only, and attempts to cultivate them lower down have not met with success. These include the three 'Silver Mimetes', and the 'Marsh rose', *Orothamnus zeyheri*, which is almost extinct. Once it was missing for 90 years until it was found again in 1913.

The mountains at Franschhoek. This is the same range extending to the north. The little 'Blushing bride', *Serruria florida*, which has several times been believed to be extinct, once for as long as 100 years, can still be found high in these mountains. Now it is popular in gardens, having been preserved by the National Botanic Gardens at Kirstenbosch.

The rugged Swartberg is the natural habitat of several rare and interesting species of Proteaceae.

This range lies between the arid Great Karoo and the Little Karoo, but snowfalls are not uncommon in winter.

Examples of Australian Proteaceae

Telopea speciosissima

The Waratah is about the only exotic species which is easily recognized by the layman as being a relative of our protea as it has coloured bracts. The sturdy tree is about 4 metres high. It is becoming popular in South African gardens.

Grevillia banksii

The flowerhead is reminiscent of that of our leucospermums. This is a handsome small tree about 3 metres in height, and is becoming well known in South Africa. Here it flowers almost throughout the year near the coast. The genus Grevillia is a large one with 250 species, most of them small trees or shrubs but the Silvery oak, *Grevillia robusta*, grows to a height of about 10 metres.

Hakea sericea

was first introduced into this country as a suitable hedge plant. Now our mountains are infested with it and the prickly shrubs threaten to choke out our own Proteaceae.

How to Grow Proteas

It comes as a surprise to many people to hear that proteas were first successfully cultivated in England and Europe. The days of the sailing ships were the days of the explorers, and the members of an expedition usually included a person who had some botanical knowledge, and who brought home dry specimens, seeds, and even young plants.

It was fashionable to grow exotic plants in the gardens of the nobility. The hot-houses were heated by means of warm air and it was only with the cult of the exotic orchid that steam-heating was introduced. This proved fatal for proteas, which cannot tolerate close, humid conditions, and the art of growing them was lost to Europe.

Here at the Cape the majority of gardeners remained sadly indifferent to the many indigenous species our own country has to offer, being under the impression that our wild flowers are difficult to grow away from their natural habitat, or that the flowering period is too short.

Now the latter may be true of many kinds but it does not apply to the most admired of all, the proteas, which flower for three months or longer. In most species the flowering period is in the colder months of the year, and they still provide occasional blooms long after the flowering period is over. A protea garden could, in fact, provide you with flowers throughout the year (but where they are cultivated in a different climate they may have a different flowering time).

That they are difficult to grow in gardens is a mistaken idea which was corrected by Marie Murray Vogts when her book, *Proteas, Know Them and Grow Them*, was published in 1958. Mrs. Vogts had spent twenty years studying the physiological requirements of the various species of Proteaceae. As a result more and more proteas are being successfully grown in gardens today. Once you understand three or four of their essential requirements, proteas are easy to grow and, once established, need little attention.

Probably the commonest mistake the early gardeners made was to plant them in their flower-beds. Although proteas have deep roots, they also have a fine network of hair-roots, just below the surface of the soil, which should on no account be disturbed. If you dig over the soil in their immediate vicinity you may kill them. They are best grown by themselves, where they can also be seen to best advantage. Allow the natural grass to cover the soil and so help to keep it cool; or you can grow a bush in the middle of your lawn provided the grass is not a deep-rooted type such as kikuyu.

It is most important that the soil should be acid—about pH 5.5. If azaleas grow well in your garden, or if the colour of your hydrangeas is blue, then you have acid soil and need not worry on this account, but it is important to keep it so. Where no proteas have been grown before it is not advisable to add manure or any fertilizer as these might make the soil too alkaline. Where lime has already been added to the soil, the resulting alkalinity can be corrected by an application of two teaspoons of aluminium sulphate to every square metre. Where the soil is naturally alkaline, one pound of sulphur to every 30 square metres is recommended.

The third important requirement for proteas is freely-circulating air. Though they can tolerate heat they cannot stand close, humid conditions. In Nature they grow best on the south-facing mountain slopes, where they thrive on the cool south-easter from the sea, so keep them away from walls and hedges and plant them in the windiest part of your garden.

Planting two or three plants close together results in a more compact bush of better shape and helps to keep the soil cool. Although young plants should be watered frequently, the soil should have good drainage and never become waterlogged.

Most proteas come into flower after the second year. Pruning is not necessary if you cut the flowers or remove the old flowerheads, leaving only those required for collecting seed.

There are now about 400 registered nurseries in the Republic where plants may be obtained, but there are, as yet, no illustrated catalogues and, with such a large variety to choose from, most gardeners are at a loss when it comes to ordering plants. The descriptions in this book will help you to choose your species from the nurseryman's list.

Should your soil be on the alkaline side it would be wisest to start with *P. obtusifolia*, which, in contrast to the vast majority of species, actually prefers a slightly alkaline soil. Should you have no success with this protea then I fear it will be no use trying any other species unless you make your soil more acid. This protea can, however, be grown in the acid soil which suits other proteas, and in this case it will be found to improve the condition of the plant if you add broken sea-shells to the surrounding soil. Other proteas which will tolerate slightly more alkaline soil are *P. susannae, P. minor, P. neriifolia, and P. aristata*.

Where the acidity of the soil is right, probably the easiest protea to cultivate is *P. cynaroides* with its large striking flowerheads on a low compact bush with attractive foliage. Others which have lovely flowerheads and neat shrubs are *P. grandiceps, P. barbigera, P. eximia, P. compacta*, and *P. repens*.

The last three species tend to become rather tall and straggly after the fifth year, but most proteas do not produce good blooms after the eighth year, when the tree should be replaced, even though the life might be about thirty years.

If you wish to collect seed, the flowerhead must be picked only after it has become dry on the plant. It can then be placed in a muslin bag and hung up in a dry and cool place until the seeds become loose and are easily removed.

The seeds can be planted in seed-boxes in the usual way or they can even be planted *in situ*. Not a large number of the seeds will germinate and the process will take between two and five months. You will have more success in the colder months, preferably in late autumn. Young plants should be protected against frost.

It is of interest to me that protea seeds appear to have some built-in conditioning which prevents them from all germinating at the same time. Even if you take seeds from the same flowerhead and sow them in the same seed-box so that conditions are uniform for all, then you will find some germinating within a few weeks and others will follow at varying intervals which may be more than five months. This is obviously one of Nature's little miracles to ensure that at least some of the young plants will have a good chance of experiencing favourable weather conditions over the first few critical months.

It is well worth while becoming a member of the Botanic Society which has its headquarters at Kirstenbosch. The annual subscription is only a small amount, but members have the privilege of being able to order seeds of wild flowers from the National Botanic Gardens at no cost.

The principles given above apply to most species of Proteaceae and anyone who has a bit of vacant land from which he wishes to reap the maximum of pleasure with the minimum of care should consider establishing a protea garden. You can find something exciting there on any day.

Proteas Lost—and Found Again

Several of our rare and most fascinating species have been believed to have become extinct, fortunately to be rediscovered later, often after so many years that all hope had been lost.

Protea odorata (page 31) was first described by Andrews from a plant cultivated in England, but here at the Cape it was believed to be extinct. It was thought to have come from the Cedarberg but could never be found there.

This is the protea with the smallest flowerhead—measuring only about 3 cm across—and it was also believed to be the only protea to possess fragrance, so it was with considerable excitement that Prof. H. B. Rycroft, of the National Botanic Gardens at Kirstenbosch, received a report in 1954 that there were a few plants growing not far from Cape Town at Fisantekraal. The description tallied with that given by Andrews 150 years earlier, so you can imagine Rycroft's dismay when he visited the area only to find that an aerodrome was under construction and all the plants exterminated by extensive earth-moving.

However, shortly afterwards Prof. Krausel of Germany sent a specimen to the Herbarium for identification, reporting that he had found it in the veld between Philadelphia and Malmesbury. Prof. Rycroft set out to comb this large area and, to his joy, he found a specimen near the road at Kalbaskraal. He asked a shepherd living near by to look after the plant carefully, explaining that it was the only one of its kind in the world. This statement brought the remark, 'My mastig! Dan het die baas darem al baie ver gery.' (Goodness! Then the master has travelled very far.)

This plant is now preserved at Kirstenbosch. Another of our beautiful proteas which Prof. Rycroft feared we had lost was *Protea grandiceps* (page 16). It was then known to grow only in the mountains at Jonkershoek, near Stellenbosch, where Rycroft was then a Forestry Research Officer. Fire destroyed all the plants except one and this he nursed very carefully. He exhibited one of the flowerheads at the Stellenbosch Horticultural Show in 1952 and later, when he returned to gather seed, he was dismayed to find that vandals had stripped this solitary plant of its remaining six flowerheads.

Fortunately he found that Mr. Frank Batchelor, of the farm 'Protea Heights', had gathered seed before the fire and had not only succeeded in getting them to germinate, but had hundreds of young plants. Today *P. grandiceps* is popular in gardens. It is not extinct in Nature as it has since been found high in the mountains at a few other spots, but it is rare.

The little 'Blushing bride', *Serruria florida* (page 63), was first discovered in the Franschhoek mountains by Thunberg, and the story goes that, in the days of the early settlers, if a young man courted a girl and he wore one of these little flowers in his buttonhole then it meant that his intentions were honourable and he had matrimony in mind. This caused the maiden to blush very prettily. The 'Blushing bride' was rare even in those days and grew high in the Franschhoek mountains only, where it was soon destroyed by fire, and for 100 years it was believed to be extinct, until 1891, when Professor Macowan visited a wild-flower show at Franschhoek and could hardly believe his eyes when he recognized the 'Blushing bride' among the exhibits. Since that time it has been feared to have become extinct more than once, but there are still some plants high in the mountains. It has been preserved at the National Botanic Gardens at Kirstenbosch and is popular in gardens today.

Protea aristata (page 35) was first discovered by Stokoe in 1928 but it was only named in a published description by Phillips in 1938. It could never be found again and was believed to be extinct until twenty-five years later when it was rediscovered in the Klein Swartberg near Ladismith and, of all places, in the Seweweekspoort near a popular camping-site near the road and, of all times, flowering just before Christmas. It is of interest because the small tree, only about 1 metre in height, has needle-like leaves and looks like a small pine tree. Proteas with such narrow leaves are classified in the section Pinifoliae, and have small rose-shaped flowerheads, but *P. aristata* has the large cup-shaped flowerhead and spoon-shaped bracts of the proteas which fall in the section Ligulatae, and these all have broad leaves. The bracts are deep red in colour and the red flowers are attractively white-tipped, giving them the appearance as if they have a fine coating of powdery snow.

There were five trees growing close together in the Seweweekspoort and I concluded that they must have come from seeds, either in a bird's nest or in a dry flowerhead, which must have blown down from higher in the mountains. Subsequently more trees were found much higher up. It has been cultivated and preserved at the National Botanic Gardens at Kirstenbosch and appears to respond well to cultivation, preferring a slightly alkaline soil.

I have come across this lovely protea several times and only later did it occur to me that, on each occasion, I was aware of a faint but distinctly unpleasant odour and, on each occasion, I had examined the soles of my shoes and found nothing there to account for it.

The most outstanding member of the family is the rare and exquisite Marsh Rose, *Orothamnus zeyheri* (page 47), which is at present threatened with extinction as it has not responded to cultivation. There has been some success in getting the seeds to germinate, but the young plants soon damp off, and it evidently needs the misty conditions of the high mountains.

It was first discovered by Carl Zeyher more than a century ago but it could never be found again and was believed to be extinct until ninety years later when, in 1913, the well-known wild-flower photographer, F. J. Steer, bought two from a flower-seller in Adderley Street. When these were identified, W. H. Paterson was determined to find them and started a long search which finally resulted in the discovery of some plants high in the mountains above Hermanus. It has subsequently also been found in one or two other spots high in the neighbouring mountains, but there are very few plants left. Vandals have taken their toll because, if you pick the long stem below the leaves, the whole plant dies. (The lower part of the stem is devoid of leaves.) Some suspect that the plant may have been overprotected and that an occasional fire might be beneficial. This would have the effect of clearing away dense vegetation, thereby enabling seeds to germinate and young plants to establish themselves. Experiments are now being conducted along these lines by the Department of Forestry.

Common Names

Probably because not many species are familiar, comparatively few have popular names. These names tend to be confusing since the same common name is sometimes given to more than one species. On the other hand, the same species may be known by a different name in another locality. The name, sugarbush, is generally applied to any species in the Genus Protea, but is specifically confined to *P. repens* in certain areas where this particular species is common. In the following index the Afrikaans equivalents appear in brackets next to the English common names.

Common Name	Species	Page No.
Bearded proteas (baardproteas)	*Speciosae*	12
Bergkatjiepiering	*L. humifusum*	80
Blushing bride (bergbruidjie)	*S. florida*	63
Bot-river protea	*P. compacta*	11
Christmas protea	*P. aristata*	34
Giant protea (reuseprotea)	*P. cynaroides*	3
Golden tips (geelbos)	*L. adscendens*	74
Golden tips (geelbos)	*L. xanthoconus*	75
Ground protea (grondprotea)	Underground stem	40–46
King protea	*P. cynaroides*	3
Marsh rose (vleiroos)	*O. zeyheri*	47
Mountain rose (bergroos)	*P. pityphylla*	30
Mountain rose (bergroos)	*P. witzenbergiana*	30
Mountain rose (bergroos)	*P. harmeri*	32
Mountain rose (bergroos)	*P. cedromontana*	33
Mountain rose (bergroos)	*P. pendula*	33
Perdekop	*L. reflexum*	53
Pin-cushion (speldekussing, luisie)	*Leucospermum*	48
Pride of (prag van) Tulbagh	*P. barbigera*	19
Queen protea	*P. barbigera*	19
Rooibos	*L. adscendens* (Langkloof)	74
Rooistompie	*M. cuculata*	60
Silver tree (silwerboom)	*L. argenteum*	92
Skaamroos	*P. sulphurea*	36

		Page No.
Skaamrosie	*P. nana*	30
Snow flower (sneeublom)	*P. cryophila*	46
Stinkblaarprotea	*P. susannae*	9
Sugar-bush (suikerbos)	*P. repens*	6
Tolbos	*Leucadendron*	71
Waboom	*P. aborea*	38
Wild almond (wilde amandel)	*Brabeium stellatifolium*	70
Woolly-bearded protea (wolbaard)	*P. barbigera*	19

Index

	Page
AULAX	68
A. cneorifolia	68
A. pallasia	69
A. pinifolia	69
AUSTRALIAN PROTEACEAE	102
Grevillia banksii	102
Grevillia robusta	102
Hakea sericea	vi, 102
Telopea speciocissima	xiii, 102
BRABEIUM	70
B. stellatifolium	70
DIASTELLA	97
D. serpyllifolia	97
FAUREA	97
F. macnaughtonii	97
F. saligna	97
LEUCADENDRON	71
L. aemulum	81
L. adscendens	74
L. album	93
L. argenteum	92
L. aurantiacum	93
L. chamelaea	85
L. comosum	81
L. conicum	93
L. conocarpum	50
L. cordatum	80
L. daphnoides	79
L. decorum	73
L. decurrens	85
L. discolor	87, 88
L. dregei	90, 91
L. elimense	86
L. eucalyptifolium	82
L. galpinii	94
L. gandogerii	72
L. glabrum	85
L. glaucescens	88
L. grandiflorum	76
L. guthrii	72
L. humifusum	80
L. loranthifolium	88
L. laureolum	73
L. microcephalum	71
L. macowanii	84
L. modestum	86

	Page
L. nervosum	83
L. pearsonii	88
L. platyspermum	83
L. plumosum	89
L. procerum	88
L. pubescens	95
L. salicifolium	82
L. salignum	74
L. saxatile	72
L. sessile	78
L. spissifolium	85
L. stokoei	71
L. strictum	82
L. strobilinum	72
L. tinctum	76, 78, 80
L. tortum	94
L. uliginosum	84
L. venosum	78
L. xanthoconus	75
LEUCOSPERMUM	48
L. album	59
L. attenuatum	49
L. bolusii	51, 59
L. buxifolium	56
L. calligerum	56, 57
L. candicans	59
L. catherinae	54, 55, 59
L. conocarpodendron	50
L. cordifolium	51
L. crinitum	57
L. cuneiforme	51, 56
L. glabrum	50
L. grandiflorum	55
L. gueinzii	55
L. hypophyllocarpodendron	59
L. hypophyllum	59
L. incisum	50
L. lineare	53
L. mundii	57
L. muirii	56
L. nutans	51
L. oleaefolium	57
L. patersonii	51
L. prostratum	58
L. puberum	57
L. reflexum	53
L. rodolentum	59
L. tottum	52
L. spathulatum	48

	Page
L. truncatulum	56
L. vesititum	50, 51
MIMETES	60
M. argentea	62
M. capitulata	62
M. cucullata	60
M. hartogii	60
M. hirta	61
M. hottentotica	62
M. lyrigera	60
M. pauciflora	61
OROTHAMNUS	47
O. zeyheri	47, 100, 106
PARANOMUS	66
P. bracteolaris	67
P. crithmifolius	67
P. sceptrum—gustavianum	66, 67
P. spicatus	67
P. reflexus	67
PROTEA	3
P. aborea	38
P. acaulis	42
P. acaulis var. *cockscombensis*	42
P. acerosa	43
P. amplexicaulis	43
P. aristata	34, 104, 106
P. aspera	40
P. barbigera	19, 20, 21, 22, 23, 24, 25, 104
P. caespitosa	44
P. caffra	39
P. canaliculata	31
P. cedromontana	33
P. compacta	11, 104
P. comptonii	39
P. convexa	46
P. cordata	45
P. cryophila	46
P. cynaroides	xv, 3, 4, 5, 19, 104
P. decurrens	44
P. dykei	38
P. effusa	35
P. eximia	8, 104
P. gaugedii	39
P. glabra	26, 29, 39
P. glaucophylla	43
P. grandiceps	16, 104, 105
P. grandiflora	38
P. harmeri	32
P. hirta	39
P. humiflora	43
P. incompta	18
P. lacticolor	28
P. lanceolata	29
P. latifolia	8
P. laurifolia	13
P. lepidocarpodendron	13
P. longiflora	27
P. longifolia	10

	Page
P. lorea	41
P. lorifolia	10
P. macrocephala	18
P. marlothii	35
P. mellifera	6
P. minor	34, 104
P. montana	45
P. multibracteata	39
P. mundii	27
P. nana	30
P. neriifolia	12, 104
P. obtusifolia	9, 17, 104
P. odorata	31, 105
P. patersonii	14
P. pendula	33
P. pityphylla	30
P. pulchra	17
P. punctata	26
P. recondita	35
P. repens	6, 17, 104, 107
P. restionifolia	45
P. rhodantha	39
P. rosacea	30
P. rouppelliae	10
P. rubropilosa	39
P. rupicola	38
P. scabra	40
P. scolopendrium	44
P. scolymocephala	32
P. simplex	39
P. speciosa	14
P. speciosa var. *augustata*	14
P. stokoei	15
P. subpulchella	17
P. sulphurea	36
P. susannae	9, 104
P. tenuifolia	40
P. transvaalensis	39
P. tugwelliae	45
P. turbiniflora	44
P. venusta	37
P. welwitchii	39
P. witzenbergiana	30
SERRURIA	63
S. aemula	64
S. adscendens	65
S. artemesiaefolia	65
S. barbigera	64
S. burmanii	65
S. elongata	64
S. florida	63, 100, 105
S. pedunculata	65
S. rosea	64
SOROCEPHALUS	96
S. imbricatus	96
SPATALLA	96
S. caudata	96
S. setacea	96